Magic Water : The Secrets of Alkaline Water and a Long Healthy Life

by Jerry Dukes

ISBN 978-1-300-13744-3

CONTENTS

Chapter 1   What is alkaline water? What does it do? Ionizers and drops. Too much alkalinity? Ageing and reverse ageing.

Chapter 2   Cells are immortal, or should be. What has been established?

Chapter 3   Diet. Acidic food. Nutritional deficiency disease. Science of foods. Alkaline food. Recommended diet

Chapter 4   Cancer. Viruses and cancer.

Chapter 5   Exercise. Weight loss. Summary.

Chapter 6   The politics of medicine.

Chapter 7   The principle method.

Chapter 8   More about acid and alkaline water. Killing bacteria and viruses. Human skin. Cooking. Whisky.

Chapter 9   Far infrared (FIR) technology. What is it? Heating. Heaters. Saunas. Cooking. Waterless egg boiler. Healing.

Chapter 10   Other theories of ageing.

CONCLUSION. Cheers!

APPENDIX..

END NOTES/BIBLIOGRAPHY ....

Disclaimer

This book is designed to educate. It is sold with the understanding that the publisher and author shall have neither liability nor responsibility for any injury caused or alleged to be caused by the information contained. If you feel the need, please contact a medical professional for any illness or disease. The FDA of America has made no review of ionized water and therefore has made no determination about it. Ionisers are approved by the Japanese Government as a health therapy.

Author's email address : jerry.dukes@bigpond.com

INTRODUCTION

I was forty when my dad died suddenly from a heart attack at age 73. A successful businessman, he was healthy all his life. My mother died at age 93. Soon after Dad died I knew I could have saved him. Involved in a health and fitness regimen which focused on the eating of more vegetables, more fruit and less animal fat, I saw that if he had just watched his diet a little more carefully, he too might have lived to his nineties.

Unfortunately, the evidence was clear, his regular diet rich in saturated fat and cholesterol had blocked up the arteries and, sadly, his first heart attack was his last.

I had become interested in the connection between nutrition and health as a child. My mother had discovered The Hay Diet and insisted that the family adopt it. It was a diet that separated foods into starches (carbohydrates) and proteins which were not to be eaten together in the same meal. The starches were mainly foods like potatoes, breads and cereals; the proteins comprised meat, fish, eggs and animal foods. Vegetables and fruits were neutral and could be eaten with either starches or proteins. Mum said the starches and proteins could not be eaten together because the stomach could not digest them properly.

So... for lunch we had starches, and for dinner we ate proteins. We hated it! Dad and his four kids didn't like their meat without the potatoes or vegies without meat. But as we were all doing so well in the health department, Mum was convinced things should stay that way. Like most diets, The Hay Diet served its time; and when contrary scientific evidence appeared in the media it was forgotten, that is, until the next popular diet arrived. But it made me think.

After university and considerable reading over the years I concluded that vegetables, fruits, nuts, meat and fish were the main natural foods for humans. Over the ages, fruits and vegetables would have provided the bulk of the energy with nuts, meat, and fish making up lesser amounts. A million years on this diet had made us what we are now or at least what we were when we stopped being hunter-gatherers and became farmers about 20,000 years ago. This was when we started to grow crops of cereals and switched to a much higher carbohydrate diet. This, it has turned out, was going against genes that had developed in our bodies for eons : it caused a shift in our health and wellbeing.

Potatoes, the other main carbohydrate with cereals, may not have been popular in prehistoric times because potato leaves were poisonous, and digesting the raw underground potato was a tough ask. Mr Stone Age man would probably have condemned the whole plant and left the tuber in the ground. It follows that carbohydrates were very likely a very small part of human diet up to 20,000 years ago when suddenly cereals and perhaps potatoes became the main diet. The concensus among present-day researchers is that 20,000 years is not long enough for a change in our genetic make-up. Excess quantities of carbohydrates since that time have upset our metabolism and affected our health in a negative way. More about this in Chapter 3.

*

Next was the Atkins Diet. In my early forties, I had my own import/export business. Married with three children, I was eating and drinking too much, did little exercise and got fat. It was time to take stock. The Atkins diet certainly took off the weight. It allowed a large intake of proteins (meat, fish, nuts) and also fats but severely restricted carbohydrates (cereals, potatoes, sugars). Most vegetables and fruits were allowed as neutral. In a year I went from 80kg to 60kg and had to buy new clothes.

But something was wrong!

My cholesterol had gone up alarmingly. The doctor said I had to cut out the fats and oils, especially saturated fats, or risk a heart attack. Goodbye Atkins Diet.

Then along came the Pritikin Diet. I read *The Health Revolution* by Ross Horne who had studied the books of Nathan Pritikin. These pioneer authors told us how we needed carbohydrates for energy or the body would burn off fat instead, leaving unwanted chemicals in the blood. It was completely opposite to the philosophy of Atkins. The media in America arranged a television debate between diet gurus Pritikin and Atkins. The result? The only thing they could agree on was that Americans were too fat. Atkins stressed he was a doctor and that Pritikin was not. Atkins was a GP; Pritikin was a scientist, an engineer/inventor with a keen analytical mind.

Along the way I read Professor Roy Walford who introduced his restricted calorie diet for longevity.[1] He proved that animals and insects fed 40 percent fewer kilojoules (calories) than normal with highly nutritious food could double their lifespans, and he presumed humans could do the same. This doubling of lifespan applied only to those starting very young : the old and middle-aged would have a shorter increase in lifespan than the young. It seemed a tough ask to eat only highly nutritious food and a lot less of it. I put it on the backburner.

But what an amazing change the Pritikin diet made for me back there in my forties! After starving my body of carbohydrates for twelve months on the Atkins Diet I suddenly felt I had the energy to just about jump over the moon! I was living again, interested in life and people; so much so that I spent the next 18 years writing a monthly eight-page newsletter for the Pritikin people which went out to 2,000 readers around Australia.

The Health Department said my newsletter, which contained extracts of medical research reports from all over the world, was well received in their office. They said it helped them in their own research for a new nutritional health policy for Australians. That policy, very close to Pritikin's low fat, high complex carbohydrate, high fibre, and low salt guidelines, is still the official dietary recommendation for Australia, the UK, and America and also my habitual diet for the last twenty-five years.

Although told by doctors I had the body of someone twenty years younger, I had begun to sense a slow but gradual decline towards old age. I realized that if I wanted to live in good health for many years to come, I needed something more: I needed a better ageing method.

And I believe I found it.

My quest for staying healthy and young lay simply in "Magic Water."

*

Scientist Sang Whang of Florida U.S. may have first used the term 'magic water' to describe alkaline water in his 1990 book *Reverse Aging*[2]. Whang, a Korean-American, followed this book seventeen years later with his *Aging & Reverse Aging*[3]. These unpretentious little texts held a compelling message. They described how the body accumulated toxic (poisonous) acids during a lifetime of poor eating and also through the natural process of ageing. The Japanese, who are the longest living industrialised people in the world, have been consuming the so-called magic water, or alkaline water, for more than sixty years.

All you needed to do to lose weight, retard or stop ageing, or even reverse ageing, Whang said, was to drink alkaline water made by filtering ordinary tap water through a domestic ionizer or alternatively adding a few drops of the patented chemicals potassium hydroxide and sodium hydroxide in the right proportion to tap water. No change in diet needed, no extra exercise required, perfect health and an extra long life practically guaranteed. It all seemed too good to be true. I bought an ionizer and drank alkaline water for the next four years, and I'm still counting.

Whang was a scientist, someone who dealt in atoms, molecules, chemical equations. He wasn't a doctor, a nutritionist, or a dietician. But with Sang Whang the science of anti-ageing in the West suddenly took a stunning new turn.

*

This book is about getting rid of your illnesses. It is also about allowing you to live a very long time in good health. It is even about making your body as old as it was, say, 10, 20, or even 30 years ago. It will, or should, prevent you from diseases such as cancer, heart and respiratory troubles and diabetes to name a few. If that sounds a bit far-fetched, let's just say the book is about making you healthy, even though I believe it is those other things also.

The main cause of degenerative diseases presently plaguing our lives: heart disease, cancer, dementia, lung, blood and digestive diseases and about sixty other body diseases can be pin-pointed to one thing: a build-up of acid in the body. Very few medical professionals are even aware of this simplistic fact.

The medical profession is very knowledgeable about infectious diseases: those illnesses and deaths caused by the invasion of viruses and bacteria. The medical record for finding and beating germs is excellent. Infectious diseases are being, or have been, eliminated one by one all around the globe. But the build-up of unwanted acid wastes in the body is obviously not a subject the medics have been made aware of, at least in Western countries.

We all know that acids neutralise alkalines and that alkalines neutralize acids. Acid and alkaline chemicals are given a pH number. (pH means "potential hydrogen"). Acids are pH 1-6; neutral is 7; alkalines 8-14.

The medics know there is acid in the body. There is acid in everyone's blood, in bones, muscles, tissues, and kidney stones as well as fat cells which make us fat. Because acid is everywhere in the body, it seems doctors and people apparently believe it is there because it needs to be there. Why else would it be there, they might ask? They also know that acid wastes build up because of age : the older you get, the more chances you have of having acid in your fat cells, your kidney stones, muscles, bones, blood, tissues, etc.

Do you get acid wastes because you get old? Whang's message is : *No, you get old because you get acid wastes.* You also get sick and unhealthy. If you can eliminate all acid wastes in your body as they form daily and also those which have built up over the years, you can be relatively free of disease and live a long time in good health. You can even become physically younger. This book is based on simple facts; you don't need to be an Einstein to understand it. Read on.

CHAPTER 1

WHAT IS ALKALINE WATER?
Pure mountain stream water is highly alkaline, in fact it has a pH of 10. Whang told of the time he visited a resort in the mountains of Colorado when he took his ionizer with him. When he tried to ionize the water from a nearby small mountain stream which originated from a completely natural, non-human environment, he found the machine could not produce acid water, only strong alkaline with a pH of 10.8. The water from the ionizer's acid outlet was 7.0, quite neutral. To Whang it seemed unpolluted alkaline mountain stream water was what our bodies had evolved on and what we needed for good health and longevity.

When a Japanese soldier in the Philipines at the end of WW2 saw that the Americans would soon defeat his army and he feared torture and death from the enemy (as he had been taught to believe), he took his rifle and not much else and went up into the mountains. He stayed there for 35 years before emerging to give himself up to the authorities. He was shocked to hear the war had ended long before. Upon checking him out, doctors found to their amazement he still had the body and organs of a young man. Despite all his hardships he had not aged. He had lived high above the mosquitos and diseases, drinking only pure mountain water. Alkaline water!

I have consistently imbibed alkaline water with a pH of 9.5 for the last four years and have never felt better : no aches or pains, no illnesses, a brain sharper than ever, and a great feeling of wellbeing 24/7.

City tap water can be split into acid water and alkaline water by passing it through a small electrical filter which usually contains two electrodes, one positive and one negative. Some models have from three to seven pairs of electrodes. The filter is called an alkaline filter or ioniser, and it splits the water according to its mineral content which is either mainly alkaline or mainly acid. The mainly alkaline water comes out through one pipe with a maximum pH of 10.3 and the mainly acid water through another at pH3.8. Alkaline water is for drinking and cooking, and acid water is for washing and protecting the skin and hair. One for the insides and one for the outsides. (More in Chapter 8.)

*

Pure water has a pH of 7 and is neutral. When pure water absorbs minerals from the soil, pollution, or by travelling over various rock formations, it changes the pH depending on whether the minerals are mainly acid or mainly alkaline. Tap water usually contains both, i.e. it contains a mixture of alkaline and acid minerals. Generally it has a pH of about 7.5 or 8. The alkaline minerals are potassium, magnesium, sodium, calcium and iron. The acid elements are sulphur, phosphorus, chlorine, iodine and urates.

*

WHAT DOES ALKALINE WATER DO?

In a nutshell, alkaline water produces extra bicarbonate for the blood which neutralizes and expels built-up acids. Here is how alkaline water works:

When alkaline water goes into the stomach it neutralizes the hydrochloric acid therein, increasing the acid's pH and making it ineffectual for digestion. The stomach immediately takes carbon dioxide from the lungs and sodium salt from the blood--which seeps through the stomach wall--and with water turns the mixture into more hydrochloric acid on an instant needs basis. It is produced on an instant needs basis because if acid with a pH of less than 4 was stored in the stomach it would burn a hole in the wall. This newly-produced hydrochloric acid lowers the pH of the stomach to normal again, about pH 4, so it can do its job of digesting the food. The other by-product from this mix is sodium bicarbonate. (Note : it is *bi*carbonate not carbonate.) It is the bicarbonates we are interested in - they are the magicians who wave their magic wands for immortality.

*

Here is the chemical equation: $H_2O + CO_2 + NaCl = HCl + NaHCO_3$.

If you want a beginner's explanation of this equation, see the Appendix. It tells you how water plus carbon dioxide plus sodium salt are turned into hydrochloric acid and sodium bicarbonate in the stomach.

Note this important fact:

The stomach cannot produce hydrochloric acid without producing sodium bicarbonate. Alkaline water swallowed forces the stomach to produce hydrochloric acid. The more hydrochloric acid the stomach makes the more sodium bicarbonate it makes. The acid goes to the stomach; the bicarbonate goes into the blood. It is the quantity of bicarbonate in the body and blood that determines whether or not we live to a very long age in good health. The more, the better. Bicarbonates can be potassium bicarbonate, sodium bicarbonate, calcium bicarbonate, or magnesium bicarbonate.

When food leaves the stomach and goes to the small intestine it is very acidic and could damage the intestine. The pancreas (place your hand just above your belly button and your pancreas is there) produces pancreatic juice which is alkaline and which neutralises the acid. In creating this alkaline juice, which is sodium bicarbonate, the pancreas must also create hydrochloric acid which enters the bloodstream. Whang explains post-meal dozing:

> *We experience sleepiness after a big meal--not during the meal or while the food is being digested in the stomach--but when the digested food is leaving the stomach. Hydrochloric acid is the main ingredient in antihistamines that cause drowsiness. Alkaline or acid produced by the body must have an equal and opposite quantity of acid and alkaline in the blood; therefore there is no net gain. However, alkaline supplied from outside the body, like alkaline water, results in a net gain of alkalinity in the body."* [4]

*

We have about 75 billion cells in our body and each one has a small furnace inside it called the mitochrondrion. This little fireplace burns carbon (food) and oxygen to produce energy, without which the body will die. All fires produce an ash of some sort, and the ash from the cell is acidic (i.e. almost totally acid). Bicarbonate in the blood withdraws the ash from the cell, neutralizes it and expels it from the body mainly by way of perspiration but also by urine.

At least, that is what is supposed to happen.

The trouble is, not all this ash waste is expelled. In this modern day and age and as a result of bad diet, stress, and lack of exercise, we don't produce enough bicarbonate. Also, bicarbonate production in the body diminishes with age. Here is a graph by Dr

Lynda Frassetto showing how bicarbonate is lost from the body with age.

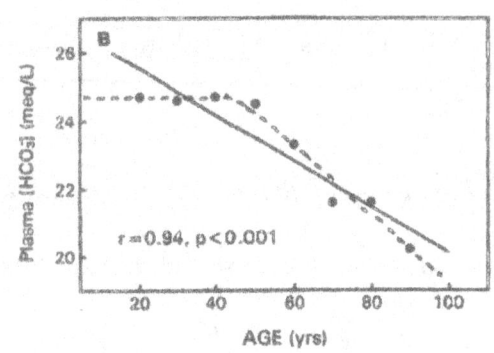

Frassetto Graph

Dr Prakova of Bulgaria studied 100 workers in a polluted factory environment and compared them with other workers in a cleaner position, the office. Ten years later he saw a noticable decline of bicarbonate content in the blood of the factory people compared with that of the office people. Not only our polluted blood environment but also our modern lifestyle of special diets, daily stressful routine, and forceful "A" type personalities all contribute to a slowing of bicarbonate production in our bodies. This in turn has caused a slowing of acid expulsion. The surplus acid has remained in our bodies and has built up.

A tiny bit of acid ash waste from each of our 75 billion cells is left in the body every day. If you threw out 99.9 per cent of rubbish from your home every day and left the residue in a cupboard, over a lifetime you would have not only a full cupboard but probably also a full home. The longer you leave your acid wastes in the body, the more deep-seated they become. Even with daily alkaline water it can take quite a while for the wastes to finally disappear. It took a long time to build up the acid ; it takes time for the build-up to go. Please be patient and keep drinking alkaline water. The next section tells how to create it.

*

IONIZERS AND DROPS

Alkaline water is best created by an ionizer (aka alkaline filter), a dinky little machine that looks just like any other electrical appliance in your kitchen or like an ordinary filter. Producing water with a pH up to 10, it is usually made in Asia and distributed to Australia and elsewhere for a cost of about 1,500 Australian dollars, cheaper if bought overseas. Instead of an ionizer you can buy a special patented liquid called Alkalife, three drops of which in a cup of ordinary tap water will produce high-quality alkaline water with a pH of 10. Alkalife comes in a small bottle containing 600 drops and costs $25 delivered. The cost of drops needed for a person to drink the recommended 1.5 litres of alkaline water a day for two months is therefore $25. You have a good chance of great health and longevity for only 42 cents a day. Perhaps you would prefer an ionizer, and then the water is free.

I must add that I have no financial interest in the sale of ionizers or the supply of Alkalife drops. My interest, as it has been for all my nutritional self-publications over 30 years, is purely altruistic. I am quite healthy, look and feel thirty years younger and merely want to share the knowledge.

If $1,500 for an ionizer seems a lot, think that as you are getting at least 20 years from the machine, the total cost over that time including the occasional new filter cartridge will have been about two dollars per week. Let me tell you, the reward is worth many, many times that amount. How many of you over-forties are on a medicine that costs $30 a month and the doctor says you have to take it for the rest of your life? You just say: "That's life" and keep swallowing the pills. Insured? You still have to pay the premiums; you pay out your hard-earned one way or the other.

The promise is that if you drink five glasses (about 1.5 litres) of alkaline water every day, your body should in time become virtually free of built-up acids. You shouldn't come down with such debilitating diseases as diabetes, cancer, lung troubles, heart troubles, muscular pains, headaches, or any of about sixty other diseases of civilisation. You will, like me, have a beautiful sense of wellbeing for 24 hours of the day unless, of course, you abuse your body in some ugly fashion like smoking.

Today, the Japanese are recognised as the longest living of all industrialised peoples with average lifespan for males 78 and females 86. People in other countries die a year or two earlier on average. Many Japanese have been drinking alkaline water for 60 years. If all Japanese without exception had been drinking alkaline water for all those years, who knows what the average lifespan of Japanese would be now! As of May 2011 the oldest person in the world was Japanese man Jirouemon Kimura of Kyoto, age 114. He enjoys good health and an active mind. He worked in a post office for forty years and turned to farming at age 90. I don't know what water he drinks but I have a good guess. Mr Kimura celebrated his latest birthday with a breakfast of grilled fish, red beans, and steamed rice.

THE BUILD-UP OF ACID

The body demands that blood maintains a pH of 7.365. Any other value, higher or lower, will cause death. Too much acid therefore is fatal. In order to survive when it has too much acid and cannot get rid of it immediately, the body turns liquid acid into solid acid and stores it as kidney stones, cholesterol, fatty acids (i.e. fat), phosphates, sulphates, urates and other things in all sorts of places. It also breathes out carbon dioxide (which is acidic) from the lungs.

Therefore, when there is too much acid which the body cannot get rid of, liquid acid becomes solid acid. On the other hand, when there is too much alkaline, solid acid becomes liquid acid again. The excess acid or alkaline is neutralised by its opposite number and then expelled. These measures enable the blood to maintain a surviving pH level of, as mentioned, 7.365.

Incidently, when there is too much acid, the body robs calcium (an alkaline) from the bones; and this leads to osteoporosis giving you chalky bones. This is another way that acid is neutralised by the body before it is stored or expelled. It is also another way the body gets its own bones broken.

*

TOO MUCH ALKALINITY -- WHAT THEN?

How do you know if your body is too alkaline, that is, if by drinking alkaline water you have in the fullness of time eliminated all the stored acid wastes and have an excess of bicarbonates in the body and blood? Test your urine and/or saliva for a pH of 7 or above. If above 7, you have an excess of alkalinity. (See Appendix for details of urine test.) Excess alkalinity is relatively rare, but the following foods, which are acid-forming have a buffering effect, tend to bring the pH back to normal. Here, from the internet[5], are acid-forming foods for the occasion (which may never happen) when your body has too much alkalinity.

Slightly acid *are best,* **Most acid** *are worst.)*

Dairy and Dairy Substitutes -- Slightly acid: cow's milk, rice milk, soymilk. **Most acid**: cheese (including cottage cheese, aged cheese, and goat cheese), icecream, soy cheese, whey protein powder.
Animal meat--Moderate acid: wild fish. **Most acid**: beef, chicken, duck, eggs farmed, fish, gelatin, lobster, organ meat, pheasant, pork, poultry, seafood, squid, turkey, veal, venison.
Grains--Slightly acid--amaranth, millet. **Moderate acid**: oats, rice (brown rice, white rice), rye, wheat.

Most acid: barley, corn, rye.
Beans and legumes--**Slightly acid**: blackbeans, chick peas, kidney beans.
Vegetables--**Most acid**: mushrooms, potatoes.
Fruit--**Slightly acid**: cataloupe, dates (not dried), nectarines. **Moderate acid**: apple, apricot, banana, all berries, figs (fresh), grape, honeydew, mango, orange, papaya, peach, persimmon, pineapple, tangerine, watermelon. **Most acid**: dried fruit.
Nuts, Seeds, and Oils--**Slightly acid**: brazil nuts, flaxseeds, hazelnuts, pecans, sunflower seeds, sunflower oil, grapeseed oil. **Moderate acid**: butter, ghee, corn oil, margarine, walnuts. **Most acid**: cashews, peanuts, pistachios.
Condiments--**Moderate acid**: ketchup, mayonaise, table salt. **Most acid**: jam, mustard, soy sauce, vinegar, white sugar, aspartame, molasses, sugar cane, barley malt syrup, honey, maple syrup, brown rice syrup, yeast.

(See EndNotes/Bibliography Number 5 for Internet web address.)

*

AGEING AND REVERSE AGEING

From the foregoing it could be reckoned that the accumulation of acid causes ageing and the reduction of acid is reverse ageing; but before we can shout this from the rooftops we should clarify the statement. The decline of bicarbonates in the body is ageing and the increase of bicarbonates in the body is reverse ageing.

Both statements are correct, but : "*The decline of bicarbonates in the body is ageing and the increase of bicarbonates in the body is reverse ageing"* is more correct scientifically.

Whang said in Article 25 :

> *Bicarbonates are the primary substance for life and the basic element of nutrition. The lack of sufficient bicarbonates in our blood causes a reduction in our ability to neutralise and dump the acid. This is the cause of ageing* [6].

*

CHAPTER 2

CELLS ARE IMMORTAL, OR SHOULD BE

Cells are fed and well looked after by the blood. A hundred years ago, Dr Alexis Carrel, one of the foremost biologists of the time, took a piece of embryo chicken heart and put it in a test tube with a solution identical to the contents of a chicken's blood. He changed the solution daily. The cells grew well and kept dividing, forming additional cells. The excess cells were periodically discarded or else they would have swamped the laboratory. They kept dividing for 34 years. The experiment was finally stopped at Carrel's death when the solution was withheld and the cells died.

Similarly, human cells also divide and form other cells; the weaker or injured cell is expelled from the body. Healthy cells are intelligent enough to know that when they become too plentiful they stop reproducing. This does not apply to cancer cells which keep reproducing unrestrictedly.

Since the lifespan of a chicken is a few short years, Carrel had proved cells were immortal. It was obvious that changing the solution daily had thrown out the acid wastes and the unwanted cells from the body. Because chicken cells are much the same as those of a human, (a cell is a cell is a cell), human cells are apparently also capable of becoming immortal. The old theory says ageing causes a surplus of acid; the correct theory says a surplus of acid causes ageing.

The Japanese soldier in Chapter One obviously created the needed amount of alkaline elements in his body and blood by drinking highly alkaline mountain stream water. The resulting continuous supply of bicarbonate would have kept his blood at a safe pH and he would have had no build-up of acid in his system; no kidney stones, no fatty acids, phosphates, sulphates, or urates, and little or no ageing. He probably had small quantities of acid-forming foods and so avoided the possibility of too much alkaline.

As we mentioned in Chapter One, a surplus of alkaline in the blood giving it a pH above 7.365 is, like a surplus of acid, also fatal. We described how the body smartly solves this problem: the stored acid in kidney stones or elsewhere in the body becomes liquid again, neutralises the surplus alkaline, and sends it out of the body. The blood knows how to juggle the amounts of acid and alkaline because it contains buffers. A buffer is a chemical that cannot be changed by dilution or addition of an acid or alkaline. The buffer for alkaline is bicarbonate (HCO_3^-) and the buffer for acid is carbonic acid (H_2CO_3). These buffers juggle the quantities of acid and alkaline needed by the blood to keep the pH at

7.365.

To sumarize : the cell produces power when it burns food in the mitochondrion and leaves an acidic ash. The bicarbonate in the blood takes the acid ash out of the cell and expels it - or most of it - via the urine or perspiration. Because of our modern lifestyle, a tiny part of the residue ash stays in the cell or is made solid and stored as kidney stones, etc. When on the other hand the bicarbonate is too plentiful, the body changes solid acid stores back into liquid acid which neutralises the excess alkaline and expels it together with carbon dioxide leaving water and sodium salt.

*

SO WHAT HAS BEEN ESTABLISHED THUS FAR?
One : If we drink alkaline water, this neutralises any hydrochloric acid in the stomach raising the pH. The stomach wall secretes sodium salt from the blood which together with the water you have just given it and carbon dioxide from the lungs produces more acid and sodium bicarbonate on an as-needed basis. The stomach needs plenty of hydrochloric acid to digest the food. The more hydrochloric acid the stomach produces the more sodium bicarbonate is produced. The acid goes to the stomach, lowering the pH, and the bicarbonate goes to the blood.

Two : The mitochondria in the cells burns carbon from food and leaves an acid ash, nearly all of which is neutralized by the bicarbonate and expelled from the body. What acid is left remains in the cells or if not neutralised is stored elsewhere in the body. Any boosting of bicarbonate allows the blood to remove the residue of acid wastes from cells and built-up areas to be excreted largely by perspiration and urine.

Three : If the acid in the blood builds up to a fatal level, the body cleverly converts it into solid acids and stores it in kidney stones, cholesterol, fat, and other places as phosphates, sulphates, urates and other acids.

Four : If the blood becomes too alkaline and therefore just as fatal as too acid, the body changes the solid acids back into liquid acids which neutralise the surplus alkaline and expels it.

Five : A complete exit of unwanted acid from both built-up and newly created wastes allows the cells to become healthy and perhaps immortal. Like the chicken heart as demonstrated by Carrel, perhaps we too can become immortal.

Then again, perhaps not. So without sticking out our necks out too far, let us say that alkaline water will very likely ensure a long life with perfect health. This can still be called reverse ageing: we can reverse our health level back to where it was years ago when our acid wastes were not as high. Now that Whang has shown us the way, maybe the final piece in the puzzle of immortality is just around the corner.

*

HISTORY OF SANG WHANG

Where did Whang get all his information? When medical scientists do their research, they publish the results in regular journals. Whereas all American results usually have been translated into Japanese, Korean, and Chinese languages, only about ten percent of Asian results have traditionally been translated into English. It can be argued that Asian researchers, consequently with more information, have been more knowledgeable than American researchers, although recent disclosures have suggested a parity of this type of knowledge. Whang had hepatitus B, a fatal disease of the liver he had been fighting all his life.

Although naturally interested in improving his own health, Whang was even more interested when his doctor told him he had high blood pressure (BP) which put him at risk of a stroke. He started taking pills which reduced his BP. A woman friend, a nurse, told him that alkaline water would bring down his BP without pills. He tried this and found it worked beautifully; the BP came down to normal without pills.

But Whang, a true scientist, wanted to know just why this worked. With the help of Korean literature (he was a Korean/American) he discovered the reason and wrote his book *Reverse Aging*. He wrote it as simply as he could for the layman, but he was a professional scientist and the book contains a lot of highly technical information. A simpler book was needed. I hope it is the book you are now reading.

Whang studied the work of Professors Warburg and Koch who declared the cause of cancer was a lack of oxygen in the cell. They went a step further and said that if a cell was starved of oxygen and wanted to survive, it changed itself by using the process of fermentation. Fermentation takes in carbon dioxide and puts out oxygen; but it provides only ten percent of the oxygen that normal cells produce. Fermentation, in the opinion of the professors, was the primary cause of cancer. What caused the lack of oxygen in the cell that led to fermentation? They didn't say. Whang declared it was surplus acid. Whang may or may not have been the pioneer of this thought. Asians had been drinking alkaline water for health reasons for many years beforehand, but I am not sure if they realized the connection between acid and a lack of bicarbonate in the blood. Whang died in February 2011 before I could ask him.

But he must have had a great feeling of satisfaction when he realized this new knowledge. Archimedies made his famous eponymous law discovery about hydraulics when he was taking a bath. He raced out into the street (we hope he threw on a towel) shouting "Eureka!" Whang might have had a similar emotion when he realized his new knowledge about acid built-up and a loss of bicarbonate in the blood could make great changes in the world.

*

CHAPTER 3

DIET.   Drinking alkaline water and getting rid of the acid wastes stored in your body should guarantee you a healthy, long life; and you don't have to worry too much about acidic foods. You don't have to worry about alkaline foods either, only about getting enough strong alkaline water with a pH of 9 or 9.5. About 1.5 litres a day should be enough.

The one food we should *definitely avoid* is a fizzy drink like a cola. Fizzy drinks are very acidic. It takes 32 glasses of alkaline water to neutralize one glass of cola. Fizzy drinks also contain CO_2 which is carbon dioxide, also acidic. Plain soda water is also acidic, but if you pour it into a glass and allow most of the bubbles to come out, it is not as bad as cola. However, it is best to leave all fizzy drinks alone : there is still carbon dioxide in them and they are acidic. Think about the loss of bicarbonates from your blood. Think about the loss of calcium from your bones and osteoporosis eating away your skeleton.

*

ACIDIC FOOD

When the stomach is loaded with too much acidic food, bicarbonate may be used for neutralising that food. Instead of valuable bicarbonate being made for the blood, it is being used up in the stomach. Best to keep super-acidic foods such as mushrooms and potatoes for moderate use only. Nevertheless, with the exception of cola and fizzy drinks, acid and alkaline foods are not really the big problem. You merely need to drink alkaline water and all is well except that, in my opinion, we need nutrients just as importantly.

Getting rid of an acid waste build-up is one thing : giving the body nutrients is another. It is essential that the foods we eat are suitably nutritious. We need nutrients to survive and thrive. It's no good just getting rid of acid build-up if you are going to leave a fragile and malnourished body in its place. That is like going into battle with your body completely encased in armour but with no strength to move in it. We still need vitamins, minerals, essential amino acids, fatty acids, phytochemicals, enzymes, and several other nutrients in our diet or we leave ourselves open to illness and disease.

Most of the nutrients we need come from vegetables and fruits. Lesser amounts come from animal protein foods and dairy; most of these are still needed for health. Raw foods are healthier than cooked because nutrients are destroyed in the cooking. The one exception is tomatoes, the nutrients of which are more easily assimilated by the body when cooked. Tomato sauce is healthier than raw tomato!

Most of the diseases and illnesses suffered in our modern civilisations are caused by an acidic environment and a lack of nutrients - read bad diet. People who are usually slim and active for their age probably eat food that is more nutritious than that eaten by their bulkier peers.

Today, the one real problem facing industrialised countries including America, Europe and Australia is obesity. We are eating far too much of the wrong type of food. Nutritionists nationwide were appalled recently when a popular fast-food operator introduced a new burger with 6,000 kilojoules (1,430 calories), beating the opposition's burgers by 1,000 to 3,000kJ. We were told that one burger provided half the recommended energy for the day.

It is popularly thought that if we eat more kilojoules than we use up, we store the surplus as fat; although many nutritionists now say it is the so-called "bad" carbohydrates like sugars and cereals that cause obesity. Fat is stored as fatty acid which bicarbonates can gradually get rid of when you give alkaline water regularly to the body. However, if you make a habit of getting up from the table when you have satisfied your appetite rather than when your plate is empty you should lose weight faster.

When there is less bicarbonate in the blood, it cannot rid the cells of all acid wastes from the mitochrondrion, nor can it neutralize and expel all wastes from the body. As mentioned, the blood becomes dangerously too acidic and the body changes the liquid acid into solid acid. This is when we tend to get fat : some of the solid acid is fatty acid, commonly known as fat.

One theory of what causes this epidemic of obesity refers to food that is nutritionally poor. The theory says that when we feed ourselves nutritionally-poor food containing refined flour products and other manufactured items, the brain sends a message to the stomach to eat more food hoping that extra food will provide some of the nutrients it is missing. When the stomach is full, the brain stops sending messages with the body still needing the missing nutrients, and you remain fat.

There are those who believe in supplements like vitamin pills, etc. We don't really know if these are enough : the scientists are discovering new nutrients every day. The nutrients in the newly discovered phytochemicals in vegetables number in their thousands of types. How can today's pills cover them all? Best to eat your vegies. I personally have a bet each way : alkaline water, plenty of fruits and vegetables, most other nutritious foods (animal foods, nuts, dairy) and liquids in moderation, no pills or drugs, and no smoking.

There are many conflicting theories about food, so many so that we are all confused. With authors and their readers all different people with different needs, it seems like the blind leading the blind. Whang's scientific approach must be respected, however; his chemistry cannot be faulted.

When researchers or authorities test a drug or a procedure, they use a large number of people and a double-blind test. The double-blind test is to give half the participants the drug and the other half a placebo, a harmless substitute which looks and tastes the same. No one knows who receives what, not even the person administering the dose. Only the person running the test knows. This is to avoid the "placebo effect," someone feeling or imagining what is not real. Whang's chemical equations don't need a double-blind test, they are set in concrete by their very nature.

*

NUTRITIONAL DEFICIENCY DISEASE

It is ironical that so-called "health-conscious" people can come down with nutritional deficiency diseases. These diseases are quite new and doctors know little about them. They come from eating too much of one class of food and denying themselves other foods. This creates a nutritional deficiency. Such ill and suffering people believe they are eating only "healthy" food. Still feeling unwell, they try extra remedies like drinking reverse osmosis or distilled water (very damaging, all the minerals are taken out) thinking that "pure" water will help them. It only makes them more deficient in nutrients, in this case minerals.

Knowledge of what is good in food and drink is in its infancy at the moment, and time will rectify this. In the meantime your best bet is to partake of all your usual foods and drinks in moderation, and hold the fizzy drinks. You can try using alkaline foods to try and cleanse acidic wastes; but alkaline water will do the job much, much faster. If in doubt about food, use the advice of a professional dietitician who recommends The Balanced Diet concept. If you try alkaline foods instead of alkaline water to cleanse acidic wastes you could run the risk of nutritional unbalance or deficiency if you don't get it right. There are more nutrients in foods that we don't know about than what we do know about. The safest way is alkaline water, plenty of salad vegetables, and other nutritious foods in moderation.

*

THE SCIENCE OF FOODS

Ninety-eight per cent of food is made of carbon, hydrogen, oxygen and nitrogen atoms. Nitrogen is in protein only. Inorganic minerals both alkaline and acidic make up the other two per cent. Strangely, the above four atoms make up our earth's atmosphere. If scientists can discover how to separate, mix and combine the molecules made by these atoms correctly, they could generate food from air and so solve the problem of starvation in the world. What a challenge for the scientists! All the more reason to avoid polluting the air which we seem hell-bent on doing these days.

Whang takes issue with the health industry about promoting organic minerals as healthy for the body as against inorganic minerals which are not. He said this was the most irresponsible statement he had ever heard. Obviously, he said, it came from people who did not understand the body's chemistry and its atoms and molecules. He had the same criticism for those who promoted anti-oxidant vitamins: the industry had that all wrong, too. Whang has left papers detailing these issues in thirty-one small easy-to-read articles which he put on the Internet just before he died. I think they must have been his swan-song: he stored all his dietary and health discoveries electronically. If you have a health problem, try Whang's *31 Articles*. To locate them, search the Internet for

http://alkalife.com/page.php?&cms=article&articleid=1

The *31 Articles* are a very good and easy read and excellent for future reference. Save them in Favourites; you might be glad you did. If you don't want to type out the lengthy web address above, search for Sang Whang Official Site in Google and scroll down for *"Science & Health page"*. (Search for those words exactly or you could miss it.)

*

ALKALINE FOOD

Whang said we should not worry too much about the food we are presently eating. If we eliminate some foods and eat more of others, we could be denying ourselves necessary nutrients. His suggestion was to not overeat, eat the foods we like, and not to exclude any foods. The body needs all kinds of nutrients, and the lack of any will cause an imbalance in the diet. It follows that an imbalance in the diet leads to sickness. It also follows that eating too much nutritionally-poor food is creating your own imbalance.

DOES THE BODY TELL YOU WHAT TO EAT?

The answer to this question is perhaps clouded by the addiction factor. For instance, a smoker's body will give out the message that it needs a smoke. Is this good? No, you have an addiction. If the body is addicted to or takes in large quantities of a particular food to the exclusion of other foods, it is also questionable that more is good. But generally the body, in my humble opinion, can let you know if it is needing something necessary if you listen to it and allow for any possible addiction.

My own diet for 25 years has been one of moderate complex carbohyhdrate, moderate protein (meat, fish and lentils), low fat, high fibre, low salt, and mostly fruit and vegetable. It contains ample essential fats. I have been following this diet, or fairly close to it, for all these years and am now extremely healthy and past middle age. My conclusion is that we need such a diet in order to receive all necessary nutrients. Alkaline water takes care of the acid problem, but nutrients are necessary too and they can be obtained from a good diet. It is perhaps a coincidence that my diet for twenty-five years has been more alkaline-forming than acid-forming. I will still continue with alkaline water because acid is a continuing problem and, as Frasseto proved, the body loses bicarbonates through age.

According to animal tests by Professor Roy Walford, the nutrients which encourage health and longevity are : Vitamins C & E, BHT, selenium, some B vitamins, and several sulphur-containing amino acids.[7] Most of these nutrients come from vegetables, fruits, nuts, fish, and essential oils.

Another knowledgeable authority is American Professor Arthur De Vany who recommends we follow the diet of our ancient ancestors of a million years ago, the so-called hunter-gatherers. In his opinion our genes have not changed in regard to our diet in all that time. We now eat a diet different from that of the ancients, especially when we include cereals and potatoes. In doing so we stress our bodies in a way that leads to disease and death sooner than it otherwise would. An excess of carbohydrates can increase insulin resistance, lower testosterone, and increase oestrogen, he said, stating that couples having trouble trying to have babies might well be eating too much cereal products and potatoes.

As mentioned earlier, cereals were hardy eaten up to the age when hunter-gatherers became farmers about 20,000 years ago (some say 10,000), a mere brief period when compared with a million years and not enough time for our genes to change. If the primitive hunter-gatherers of a million years ago had eaten cereals, they would firstly have had to find the wild grains, chew them for their nutrients and spit out the indigestible grass stems. They would almost certainly have been hungry and short of other foods before they would do that, so we can't seriously imagine cereals as a significant part of their diet a million years ago. I have already mentioned potato leaves being poisonous with their underground tuber probably ignored. DeVany says carbohydrates in cereals are an unnatural food for us and the huge proportional increase of carbohydrates in our modern diet is now causing unhealthy effects.

He therefore recommends fruits, vegetables, eggs, nuts, fish, and meat. He suggests we have plenty of meat, eat the whole animal if we like, but don't eat it every day. The reason for this is that meat provides special benefits to the body and the body needs a spell from it so it can acquire the different benefits from other foods.[8] The body also needs spells of fasting as well as spells of lazing around and doing nothing in tandem with spells of intense physical effort. This, according to De Vany, is how our bodies were programmed to be lived and in his opinion they still are.

Loren Cordain promoted *The Paleo Diet* which recommended meat from grass-fed animals, fish, vegetables, fruit, and nuts being the diet of hunter-gatherers of 2.5 million years ago. Obviously agreeing with De Vany that our genes have not altered after all that time, Cordain excluded from his diet grains, dairy products, salt, refined sugars, and processed oils.[9]

De Vany and Cordain may have scored some points. De Vany and others also advocated fasting a day each week (water but no food) as quite beneficial for health, quoting that primitive man would essentially have had periods of starvation or near-starvation, and his body was, and our body now is, programmed to handle this and could actually need the occasional fast.

However, the unsatisfactory fact is we now have a multitude of authors advocating different dietary programs, some of which take one single factor and push it to its credible limits. The two main opposing dietary programs today are the low-fat and the low carbohydrate diets. Confusion most certainly reigns.

Perhaps our only option is to trust our instincts. I personally can only give you the results of four years study in university, a lifetime of reading and publishing about the health/nutrition connection, and my personal efforts which have treated me exceptionally well. It is not everyone who is told by his doctor that his body is thirty years younger than your normal Joe's. The following is the diet I have been on for twenty-five years and it has served me well.

RECOMMENDED DIET

I therefore recommend 1.5 litres (say five glasses) alkaline water a day and a nutritious diet as follows :

DAILY

* Low-fat meat, fish, or poultry - 100grams. (About the size of a cigarette pack.)

* Vegetables, all types, raw and cooked, unlimited.

* Fruit, four pieces.

* Wholegrain cereals, breads, potatoes & pastas, but in total only one-tenth the volume of vegetables and fruits combined. The scientists say the ratio of potassium to sodium is extremely important to health. Plant foods provide the correct ratio. Regarding potatoes, these are tubers rather than vegetables and are the least nutritious of the carbohydrates unless you also eat the skins which contain most of the nutrients.

* Herbs and spices to taste. The supermarkets today are full of fresh and packaged herbs and spices for cooking. Herbs such as coriander, garlic, ginger, cmin, bay leaf, dill, thyme, or lemon grass will put the lie to those who say the above mentioned foods are bland. Herbs and spices make all the difference with taste and will even get you eating more vegies instead of potatoes and bread. Just check the recipe books and you'll soon be a Masterchef.

* Salt. Use it in moderation if cooking natural foods. Manufactured foods usually contain ample salt (read the label for sodium).

ALSO

* Fish should be eaten two or three times week for essential fatty acids Omega 3 & 6.

* Vegetables are best eaten raw: cooking can and does destroy nutrients. Strangely, so does freezing: Vitamin E content is up to 80 per cent destroyed when food is frozen.

* Bread should be "wholegrain", not "wholemeal," nor white bread which is minus fibre, wheatgerm and other nutrients. Brown bread and wholemeal breads are often just white bread with colouring.

* Milk, should be skim but lite is ok in small quantity.

* Tea, coffee (not too much caffeine, decaf still tastes good)

* Fat-free, sugar-free yoghurt.

* Beans and lentils. These are very nutritious. Dr Jay Kenny of the Pritikin Longevity Centre in Florida says: "Beans, get with the beans." Kenny doesn't mention a quantity of legumes; De Vany says eat them only in moderation.

* Oils. There are good oils and bad oils. They are also called fats. The bad fats are those that are solid at room temperature and also those that go through a factory heating process which produces saturated fats, an unnatural and unhealthy food which is foreign to the body.

* Natural foods are best. Generally, the more a food is manufactured the less nutritious it is, despite the additives they put in it.

The best philosophy about diet, in my opinion, is there is no need to be religious about it. You can have some no-noes now and again; just keep them as a special treat. Try and regard the main course as one of vegetables with meat as a supplement, not the other way round. You don't need any more grass-fed animal meat than listed above; although De Vany and Cordain might disagree.

*

CHAPTER 4

CANCER

When I was about 50 and keenly interested in the connection between nutrition and health, I read in the newspaper of a potential cure for cancer. The medical researchers seemed quite confident they had the cure, but of course much more study and testing had to be done. Nothing more was heard of the cure.

A few months later there was another potential cure; and a few months later yet another one. Over the following 25 years there were regular reports of a latest potential cure for cancer, each one different from the previous one, each one needing more research and testing, and each one never heard about again, at least in the papers. It got to be a joke at the breakfast table - "In the paper today we have what must be potential cancer cure number 376 and still counting."

Cancer of course is no joke. Billions of tax dollars in the US and elsewhere are spent every year on research. Millions of people die quite nasty deaths from the disease, currently the number two killer in most countries. The great pharmaceutical companies make enormous profits from the expensive drugs used to fight cancer and the equally expensive drugs needed for the inevitable side effects as well. All drugs have side effects. *Time* magazine in October 2011 reported that the world's largest pharmaceutical company, Pfizer, earned $68 billion in annual revenue from prescription drugs.

And yet, a simple stopper for cancer seems to have been been available all along for those who have been diagnosed before the disease has gone too far. The facts are :

* Cancer cells are acidic and healthy cells are alkaline.

* Healthy cells contain full amounts of oxygen but cancer cells contain as little as ten per cent.

* Healthy cells, when attacked by acid, either die or turn into fermented cells which need much less oxygen to survive. Fermentation is the body's method of surviving an acid attack.

* Fermented cells can become cancerous. An excess of acid makes them thrive.

* Cancer cells die in an alkaline environment.

* The cause of a lack of oxygen in healthy cells is acid build-up.

These facts, with the exception of the last one, were known by scientists at least one hundred years ago. They were discussed and argued about by interested parties, but nothing concrete was done. Why, when these facts could have easily been checked out and tested in laboratories by trained scientists, was nothing done? For the life of me, I do not know. Whang did have a reason :

> It *is discouraging to see that cancer research is divided into so many specialties such as cancer of the liver, cancer of cervical cancer, cancer of the kidneys, cancer of the pancreas, etc. The cause of all cancer is a lack of oxygen caused by acidification. But when so many doctors are specialising cancer research in such narrow fields, they will never find the true overall picture. They will be forever lost in a maze.*[10]

My own guess as to why acid is not pin-pointed as the cause of a lack of oxygen in the cells is that just about everyone has acid in their cells. This may have led people to assume that acid was part of the body and it therefore needed to be there. Nothing I have read (except for Whang and Aikira, see later) has said that acid shouldn't be there. After all, present day researchers know that blood has both acid and alkaline elements and that they are juggled by buffers to give a pH of 7.365 or the body will die. Perhaps they think acid in the blood is necessary and normal, so why is acid elsewhere in the body not normal? Maybe the scientists think that acid in the body, whatever its quantity or situation, is normal.

*

History has shown the medical profession to be notoriously prone to procrastination in adopting new measures in the fight against disease. Take the case of scurvy, a disease in the 18th century that killed perhaps thousands of sailors and others who did not have access to fresh fruits and vegetables. They didn't know that such foods would have supplied them with vitamin C and protected them against the dreaded disease.

British scientist James Lind discovered Vitamin C in 1747, tested it, published it, gave talks about it and officially informed the navy that all they had to do to prevent scurvy among the seamen was to carry a quantity of limes on board for their year-long voyages. The limes would keep a long time. The sailors didn't like the taste of limes and refused to eat them. The project was dropped and the high rates of death from scurvy remained the same for the next 80 years. The navy finally made it compulsory: the ship's cooks were ordered to put lime or vegetables in food at least once a day, and the disease was stopped in its tracks. (British sailors are now called limeys by their American counterparts.)

The same story surfaced during the American Civil War when 30,000 men died from scurvy, more than were killed in action. The army doctors had been informed of the cure: any fruit or vegetable would do, but the advice was ignored and it was many years before the idea took on in America.

Another case of ignorance triumphing over truth was the incidence of beriberi in Asia. The coolies had been eating their basic food - wholegrain rice - and were fit and healthy. Someone invented a process which dehusked the rice to produce what we now call white rice. In 1884, a Japanese medical researcher, Takaki Kanmchiro, discovered that those who ate whole rice stayed healthy and many who ate white rice came down with beriberi. He told every one who would listen that there was something in the husk that prevented the disease, and he tracked it down as Vitamin B1 or thiamine. But people preferred the white rice; it tasted better and was quicker to cook. Many millions of Asians died before the appropriate ingredient was again disclosed and, forty-five years later, made widely known.

Doctors at least can be partly blamed for all these events. Doctors attend university for four years and another two as interns in a hospital. Absorbing the myriad of facts taught them by their professors who in turn were taught by their own professors, they are perhaps not to be blamed if they tend to believe the professors. It sometimes take an effort of will to change an opinion formed at school or university. The end result is that any new theory is generally looked on with some indifference or even suspicion. Perhaps this is the reason why alkaline water hasn't taken on in a large way in English-speaking countries over the past sixty years as it should have done.

The first person to tackle cancer was biochemist Dr Otto Warburg (1883-1970) who received the Nobel Prize in 1931 for his work which he detailed in *The Metabolism of Tumors*. Dr Warburg said the cause of cancer was a lack of oxygen in the cell. When a cell is faced with possible death from a lack of oxygen, the body looks for another way of getting the oxygen. The body's number one priority is survival. The only other way for the cell to do that, Warberg said, was to do what plants do : take in carbon dioxide and sugar and give out oxygen in a process called fermentation. This sometimes makes the cells abnormal (malignant), i.e. cancer forming. Malignant cells do not own a DNA memory code and are "brain damaged." They just multiply unrestrictedly and without order. But if the body's environment becomes alkaline, they can die or at least stop reproducing. Normal cells have a DNA memory code; they reproduce and grow to the level they are genetically programmed to reach and no further. In this respect normal cells are much smarter than malignant cells.

The National Cancer Institute of USA didn't verify Warburg's theories until 1950, more than twenty years later. After advising of their acceptance of the theory, it seems the institute then did nothing to determine the causes of oxygen deficiency in the human body. It was as if they said : "Nice theory, let someone else prove it." It shouldn't have been hard for scientists at that time; they had all the information they needed about atoms, molecules, chemical equations, DNA, and respiration of the cells. Why wasn't Warburg's theories tested by a large number of scientists as a matter of urgency? Another question not answered.

Another person to pinpoint the cause of cancer was Japanese medical author Herman Aihara in his book *Acid & Alkaline*. Aihara said that if the blood became too acidic, the body deposited the excess acid in cells and elsewhere in the body. This would keep the blood maintaining a healthy pH which, as mentioned, was necessary for life. The stored cells were acidic. Many cells live for four weeks and then multiply by dividing into two. The strongest cell survives, the weaker or damaged cell dies. The dead cell in turn becomes acidic and is expelled. Aihara said nothing about a lack of oxygen but said acidication was the cause of cancer. Warburg said nothing about acidication but said the cause of cancer was a lack of oxygen. Since excess acid causes the lack of oxygen in the first place, they were both actually saying the same thing.

The fact that cancer cells are acidic and healthy cells are alkaline means that cancer is the result of acidity. Healthy cells require ample oxygen, and acid restricts the oxygen needed. As mentioned, the cells turn to fermentation to provide oxygen, and they become malignant. When tumors are removed by surgery, radiation, or chemotherapy, they will begin again if the body's environment has not changed into an alkaline one. The best and probably only quick way to change body environment is to constantly drink alkaline water.

I have just finished reading *"The Emperor of All Maladies"* by Siddhartha Mukherjee which won for him the Pulitzer Prize for non-fiction writing for 2010. Stated on the book cover is : *"A Biography of Cancer."* It is an elegantly written and engrossing book; but in all its 571 pages, unfortunately, there is no mention of a build-up of acid in the body or cells as the cause of cancer or of alkaline water as a preventor of the disease. It makes no mention of the removal of surplus acid to make the body more alkaline.[11]

A Pulitzer prize-winning book will probably be read by thousands of people both doctors and non-doctors alike. Most will probably treat its contents as gospel. Those who have read it will probably reply to any question on alkaline water by saying : "It must be a fake or the Pulitzer winner would have described it." It is a shame when a truth is not passed on. When that truth should be passed on to the whole world but is not, it is a tragedy. But before we make any accusations we must first look at human nature.

What are the most treasured needs for a human being? One must be power : the power of a parent over a child so that it is brought up properly; the power of a general over his troops; the power of a manager whose corporation would go insolvent if his orders are not followed; the power of one person in a pair of people who takes the initiative while the other person follows. The list goes on. People who have power must be relatively happy in themselves; men with power consider themselves "real" men, mothers with power have pride in their well brought-up offspring. The contrary applies : those who do not have the power may resent those who deprive them of it.

Let us imagine a person who invents or discovers the secret of everlasting good health and everlasting long life. Such a person would of course have power. But think of all the people who would be deprived of power or some of it: doctors and nurses (no one is sick), pharmaceutical companies (no more need for their drugs); industrialists (no need for machinery to make drugs); all the people suddenly unemployed and therefore not fed, housed and looked after by the above companies and hospitals; and the shareholders of the company's assets whose shares are now worthless.

A crime in our society is defined as something that goes against the common good; the criminal is rightly punished. But if something that is for the common good also means that a great majority of people will feel pain and loss and that the first good is overwhelmed in importance by the second, is it a crime? The pharmaceutical people, the people who have lost their jobs, the doctors and nurses who now have no patients, would all say yes. The person who discovered the secret of eternal life and health could well be ostracised; he or she could well be hounded to death.

This has probably happened in the past. Dr Ruth Cilento wrote in her book *"Heal Cancer"* about cancer researchers including Professor William Koch who developed methods of studying oxygen in the cells. This brilliant pioneer researcher developed a synthetically produced carbonyl compound with high oxidation-reduction potential which he used very successfully as a therapeutic agent not only in cancer but in many other diseases caused by poor oxygen supply in the body. The product was called Synthetic Survival Reagent or SSR. Patients were given a small injection which under ideal conditions could destroy the free radicals which denied oxygen to the cells and allowed them to reconstitute. It put many cases of cancer into remission. Dr Cilento, with courage, wrote :

Medical therapists all over America started using SSR with excellent results. By 1940 the drug companies, especially ones making antipsychotics, had started a vendetta against Koch because SSR could also reverse dementia. I will not go into the sordid history of his destruction by the usual means of discrediting and denigrating by money-hungry, power-hungry assailants, but his product is no longer made. Koch spent the rest of his life in Brazil where, rumour has it, he was murdered.[12]

We wonder how many good inventions are sitting in the Patents Offices or company safes because suppression of such ideas has meant people will keep their jobs in inefficnt companies which will continue to make money. I suppose what is needed is a regulatory body that will allow great ideas to be released into the public arena in a carefully slow way so that people's lives are not too greatly disturbed by the transition. This would seem a more win-win solution, even though many people would do without the benefits in the beginning. But what about people who drink alkaline water being ostracized like poor Koch?

Alkaline water therapy takes some time before sure results are shown; and only if the disease, such as cancer, has not gone too far as to permanently damage the body. Our imaginary regulatory body might not be needed in such a case. Why? The cost of an ioniser or the chemical drops (see chapter 8), the delay in experiencing the beneficial results, and the inevitable critics and nay-sayers will combine to ensure a slow acceptance of the project. The lucky people will be the early alkaline water drinkers. The others may not be aware of what they are missing.

*

VIRUSES CAUSING CANCER?
Ruth Cilento says yes, viruses cause cancer, although many say it doesn't. It has been known since late 19th century. Let us say that viruses do cause cancer, but also let us say that although the cause of

cancer might be a virus, the cancer will only become stable and continue to grow if the environment is acid. If the environment is alkaline, the cancer should die along with the virus.

There have been many theories of the cause of cancer and perhaps they could all be a little right -- free radicals, lack of lecithin, lack of nutrients, insecticides, etc. But none of these would cause cancer if the environment of the body was alkaline, and the environment is made alkaline by a healthy supply of bicarbonates in the blood. Alkaline water supplies this.

What about stress? Immunologist Hans Selye wrote about stress and the immune system and how stress could cause cancer. He finished by saying : *"All research shows this : in a healthy body in a healthy environment, cancer cannot grow."* [13]

Professor Selye, like Warburg, Aihira and Kock, also probably may not have realized that a healthy environment was alkaline and not acid. How so? It would seem that because everyone they studied would have had an acid build-up and many of them were quite healthy, (even though there was a huge warehouse of the stuff hidden away in kidney stones, fat, cholesterol, etc) they probably assumed that an acid environment was as healthy as an alkaline one. Probably there were very few people around with a purely alkaline environment in order to make a comparison; so the question of acid or alkaline may not have come into it. Like the others, Selye obviously didn't know about a lack of bicarbonates in the blood that had caused the acid environment; and he probably assumed acid was the norm. As a result, Selye may not have considered the bicarbonate factor. Whang did, however.

*

CHAPTER 5

EXERCISE AND WEIGHT LOSS
The several methods for losing weight in the body are :

One : Eat less, forcing the body to burn up fatty acids which are stored as fat cells.

Two : Liposuction or surgery to remove fat.

Three : Burn up the stored fats with exercise. The fatty acid is expelled via perspiration and urine via the kidneys.

Four : Drink alkaline water to neutralize the fatty acids (fat cells) which are discharged via perspiration and the kidneys.

Methods one and two do not really work. When fat is lost from your body by eating less or by surgery it is usually replaced if and when you return to your former diet. Liposuction will also remove total fat, but it will tend to eliminate mainly surface fat and not imbedded fatty acids (fat).

Method three works after a fashion. Exercise will certainly burn up fats, but the exercise itself will create more acid. What is created is quickly neutralized, but the net loss of fat is less than if alkaline water had been used instead of exercise. When an exercise program is followed, it should be done slowly because the consequent changes in the muscles and organs of the body can cause upsets. To rush an exercise program is to risk your health.

Method four, drinking alkaline water, is the only effective and cleanest way of losing weight by reducing acids and therefore fats. Nevertheless, exercise to some degree should also be on the program. Exercise burns kilojoules, discharges acidic wastes through perspiration, and tones muscles to help build a muscular and lean body. It also warms the body and the muscles. When you over-exercise, lactic acid is built up due to a lack of oxygen which in turn is due to improper breathing or pushing beyond the limits. "No pain no gain" is wrong advice; lactic acid can cause asthma.

During exercise there are blood clots and debris in the arteries caused by acid that can pressurize the blood vessels in the brain leading to a stroke. Before starting an exercise program you should firstly get rid of clots and debris by using alkaline water. The clots and debris are acidic; the water will neutralize and expel them. Blood thinners are not the answer to the blood clot problem: they will make the blood more fluid but will not change an acid environment into an alkaline one.

An advantage of exercise is that it warms the body interior including the muscles which means that sweat gets rid of deep-seated acidic wastes and also heavy metals under the skin.

When we over-exercise, our arms and legs may be sore the following day. Stress and over-exercising will burn nutrients fast and create concentrations of acidic waste. These are created so fast that normal blood circulation cannot carry them out quickly enough, consequently they create areas of concentrated acid. When this happens, capillaries in the area get clogged up preventing normal blood circulation which would usually remove the acid. Whang adds:

Pools of concentrated acid, where nutrients are burning fast, are high in temperature. When they cool off, blood circulation is blocked. This blocked blood circulation is the cause of pain or soreness. Thermographic photos show a lower temperature at the location of pain/soreness. Since the original clogging action of the capillaries takes place at high temperature, a lower temperature shrinks the capillaries, and the removal of acidic wastes becomes more difficult. We need to heat the acid concentration area. If the soreness is not severe, we can exercise more to force the temperature up to expand the capillaries to enable blood circulation. This exercise cannot be so severe as to create more acidic wastes but just enough to heat expand the clogged up capillaries.[14]

To heat the area of acid concentration without further exercise, we can use far infrared (FIR) pads for deep heating. FIR heat penetrates deep inside the body and has a comfortable feel. For us to experience the same deep penetration by ordinary means -- using traditional heating pads -- we would have to raise the heat to a degree that the skin would burn.

Heating the area by FIR expands the clogged capillaries without creating any additional acidic wastes. When a part of the body is hotter than the surrounding parts, the body rushes blood in to lower the temperature and even it out. The blood can then carry out the acids and relieve the pain. Alkaline water, of course, creates plenty of bicarbonate which helps the acids to be expelled in the sweat and urine.

Sweat is good. Not only does it get more of the acid out of the body than urine, it also acts as anti-bacterial, anti-virus protection. The acid in sweat kills any microbe trying to get in your body. If you are not sweating or don't have dried sweat on your skin and you then cut your skin, microbes, which are with us all the time, may enter the wound and so charge into the body. We should lick the wound immediately or put saliva on it. Saliva, like sweat, is acidic and anti-bacterial. Animals instinctively know about licking their wounds. The alternative is to wash the wound with acid water from an ionizer.

*

WEIGHT LOSS SUMMARY

Exercise is one of the ingredients for a healthy, long-living body; but the trick is to take it quietly: don't overdo it or it creates more acid. As regards exercise for losing weight, gentle exercise and alkaline water will take care of that : the fat is acidic and bicarbonates will slowly, repeat slowly, eliminate surplus fat. There is a method of speeding up the elimination of deep-seated acids, clots, and debris, and that is by using far infrared heat (see Chapter 9: *Far Infrared Technology*.)

*

CHAPTER 6

THE POLITICS OF MEDICINE

When a pharmaceutical company produces a new drug that is a cure for a disease, it must be approved by a government body. In the case of the USA, that body is the Food and Drug Administration (FDA) which does a lot of the world's testing. As mentioned in Chapter 3, this body gives the drug the double-blind test. A double-blind test is to give the drug to a large number of patients suffering from the particular disease it is aiming at and to an equal number of patients a placebo (pro. pla-seebo), a dummy drugs that looks, feels, and tastes the same.

The drug must also be tested for side effects -- all drugs, without exception, have side effects --and also if there are any long-time dangers. The worst example of a drug not being safe in the long term is the infamous drug thalidomide designed for pregnant women suffering from morning sickness. It went on the market in 1957 and in 1961 was withdrawn after ten thousand human birth deformities resulted. The drug had prevented the proper growth of the foetus resulting in horrific birth defects. Obviously, the drug had not been studied or tested properly for its long term effects before approval. It is unbelievable that it took four years for the authorities to withdraw the drug, since a pregnancy lasts only nine months and deformaties are obvious at birth. All those poor little babies; the cause must have been clearly obvious; and they let the drug run for four years! It is only now in 2012 (fifty years later) that a legal class action is being taken on behalf of thousands of the victims, many of whom are minus arms, legs, or all of them. We can only hope that justice is done and the guilty are punished.

(NOTE: Alkaline water will immediately relieve morning sickness. Ref. Sang Whang, *Health & Science Article 18. Pregnancy & Aging.* [6])

Testing of new drugs by the FDA and similar authorities elsewhere in the world is expensive as advised earlier. However, the retail price of the product is adjusted accordingly and with doctors prescribing their products daily and worldwide, the multinational drug companies still make billions in profits. In October 2011 *The Australian* newspaper reported ten million scrips for drugs were written in Australia every year costing the government $580 million. Most scrips were for Lipitor, a cholesterol-lowering drug.

All this must be kept in mind by those who choose to drink alkaline water. Perhaps when you are much healthier as a result of your new water therapy you may need fewer drugs, which might make the pharmaceutical companies unhappy. Certainly, there will be objections from those with a special interest in condemning the use of the water. The cost of the ionizers or the drops/pills, the inevitable aggression and negation from some quarters will cause you doubt. Alkaline water is completely safe, it is merely doing what the body does naturally and does it better. You should make up your own minds; it's your life and your body.

Professor Hans Selye said :

> *Great progress can be made only by ideas very different from those generally accepted at the time. Unfortunately, it is literally true that the more someone sticks his neck out above the masses the more he is likely to attract the eyes of snipers.*"[15]

There is no danger in taking alkaline water, I cannot stress that enough. It can do you a lot of good and it cannot do you any harm. It increases the production of your body's bicarbonate, something that is reduced by the strains of modern life. If there is an excess of bicarbonates in your body (unlikely), a simple urine test will show this and acid-forming food will rectify it. (Perhaps even a glass or two of cola! Remember, it takes 32 glasses of alkaline water to neutralise one glass of cola !)

We are the product of millions of years of evolution, and our genes have changed little in all that time. The average lifespan of humans a million years ago was less than thirty years due to starvation, accident, predators, and disease. Death for such short-living people would not have come as the result of a lack or oversupply of bicarbonates. The ancients of a millions of years ago would very likely have had an unpolluted natural alkaline water supply with a pH of 9 or 10. Please note Whang's visit to Colorado and also the story of the Japanese soldier. A lack of bicarbonates causing disease, ageing and early death is surely a modern event.

Just as motor cars were looked on as useless before they took over from horses; just as computers were described by some as a nothing before they took over and changed the world; the boosting of bicarbonates in the blood may be ignored at first but should take over and change the world likewise. After carefully noticing during my many years a whole range of theories, therapies, and nutritional experiments in the field of medicine and also experiencing my own bodily improvements, I am convinced the boosting of bicarbonates gives the body what it was once used to over the eons, namely a high alkaline, low acid environment in body and blood and constant good health.

Bob McCauley made an interesting observation when he quoted the words of Dr. T. Baroody who said : T*hat which is built on alkalinity sustains. That which is built on acidity falls away--be it*

civilizations, human bodies, or the paper that preserves their knowledge. McCauley said :
The world's written history was recorded on alkline paper until 1850 when it began to be recorded on paper that used bleach, alum and tannin in the book-binding, all of which are acid. The records starting in 1850 are disintegrating at an alarming rate. The best that can be done is to scan them electronically and save what is left of the books by re-alkalizing the remaining paper. However, books printed on alkaline paper before 1850 still survive, often in perfect condition. Acid destroys life. A balanced, slightly alkaline pH preserves it.[16]

*

CHAPTER 7

THE PRINCIPLE METHOD

Do you ever wonder why barristers seem to enjoy their work in court more for the satisfaction of beating their opponents than whether their clients receive justice or not? I may be misjudging them but it seems they enjoy the win more because they are cleverer than their opponents in confusing the witnesses. Their tricks can include fallacies that cause the witness to blurt out something that incriminates him or makes him say something he would not normally have said.

A barrister once confided to me that he could always get the truth out of a witness and that he "did sometimes use the odd fallacy." A fallacy is something that does not qualify as an example of clear thinking. In other words, clear thinking is the absence of fallacy. Here is a fallacy : All frogs are green; my house is green; therefore my house is a frog. A fallacy. There are many different fallacies. Here is another: Lipitor is a cholesterol-lowering drug; therefore if you want to lower your cholesterol you must take Lipitor. (A fallacy: the drug may have no effect on some people; or the drug may make the patient worse; or there are other ways to reduce your cholesterol.)

The legal profession can twist words to good effect, and also, perhaps unwittingly, can the medical profession. Sang Whang spoke about this in Article 10 of his online Medical *Science & Natural Science* series. He described the method that doctors usually use in formulating a principle, and he used Sir Isaac Newton to illustrate it.

Newton was taking a nap under an apple tree when an apple fell on him. He wondered why the apple fell and so discovered the principle of gravitation, i.e. the principle that everything with mass is attracted to every other thing with mass. He also saw that if a greater force than gravity was applied to an opposite mass, that mass would fly away. That principle has enabled us to understand about flight and go to the moon. After being proved beyond question, Newton's discovery was named a principle. Principles are set in concrete.

If Newton had been a medical doctor instead of a scientist, he would have tried to make another 1000 apples fall, and when he succeeded he would say : "Apples fall." But this did not necessarily mean that pears, oranges, and other fruits also fell. He would have to repeat the same trial with 1000 of those, too. When you don't have a principle, you don't know what you are doing and must use samples and statistics. Anyone can take samples and statistics : it doesn't take imagination or any world-shattering revelations such as those that come to freakish people in rare moments. Sang Whang was such a freakish person when he discovered that good health for people can be ensured merely by boosting bicarbonates in the blood by drinking alkaline water.

Statistics, unlike principles, can be manipulated for selfish motives. This was the case with the tobacco industry which tried to avoid prosecution for damaging people's health. The industry is now being sued, still holding out with its false premises and fallacies. If you don't know what you are doing, you can also get a wrong conclusion. Whang had an example, no doubt fictitious, for that too, as follows :

Some scientists were conducting an experiment with grasshoppers. They trained 100 grasshoppers to jump whenever ordered to "Jump." After several successful tries, they continued their experiment by removing one hind leg from each grasshopper and ordered them to jump. They jumped but only short distances. The researchers then proceeded to remove the remaining hind legs. This time, no grasshopper jumped when ordered to. The researchers concluded that when you remove both hind legs, grasshoppers cannot hear.[17]

It might sound humorous, but both the legal and medical professionals have been guilty of that sort of thing. Whang said medical science relies too heavily on the sample and statistics method in fighting disease. He predicted that when medical research concentrates on the scientific principle rather than the statistical data method, the world would see a revolutionary leap in medicine.

*

CHAPTER 8

MORE ABOUT ACID AND ALKALINE WATER

If you are using an ionizer (aka alkaline filter) you will find the **acid water** quite useful. You will not get **acid water** from using the AlkaLife drops (refer Chapter 1 : *Ionizers and Drops*): you get only alkaline water from the drops. As advised, the two waters come out of an ioniser separately at the rate of about one part **acid water** to two or three parts alkaline water depending on your tap water supply.

To avoid confusion (both waters look and taste the same, which is quite pleasant), you should store the waters in separate sized screwtop glass jars. They must be glass because carbon dioxide from the atmosphere can penetrate most plastics and, being acidic, can neutralise the alkalinity of the contents. I use a glass jar (originally containing one litre of V8 juice) with coloured top for alkaline water and a different size glass jar with a different colour top for the **acid water.** Perhaps you might prefer to switch on the ionizer just when you want a drink and don't store any. Some ionizer manufacturers recommend you run the filter for a minute before using it.

To test whether this recommendation is necessary or not is to check the pH of the water. (You are given a simple pH tester with the ioniser, or you can buy one separately at any pet shop or hardware.) If the water is pH 9.5 or 10, you can use it straight away without wasting any. I fill a number of bottles at a time. Some manufacturers say to store in the frig and keep it no longer than a week, but I think the jury is still out on that one also. Alkaline water from my ioniser doesn't lose pH after a week outside the frig. It is only the pH that is important, nothing else really is, and the higher it is up to pH 10 the better and healthier for you.

Ioniser expert Bob McCauley confirms the best way to store alkaline water is in glass bottles and kept in a cool dark place. It loses its charge quickly if open to the air. Definitely, it loses it very quickly if exposed to the UV rays of the sun. The next best container is polycarbonate plastic which has the lowest potential to leach carbon dioxide from outside into the water. White plastic milk bottles have the highest potential for leaching and must be avoided. Glass is the best option.

McCauley says ionised water is strongest when fresh from the ionizer, but even if stored for a number of days it is still superior to conventional tap water. Alkaline water with high pH, say 9 to 9.5, will keep for three to eight days depending on the tap water source. **Acid water** can last up to five months if stored in a cool dark place. At a pH of 10 or above, ionised alkaline water can taste a little salty, so pH of 10 is an appropriate upper limit for drinking purposes. Ionizers have the means to adjust pH. You just press the button for strong, weak or medium alkalinity. Best to start your "habit" with weak alkalinity, going to strong over the first week or two. It's powerful stuff but you wouldn't know that when drinking it.

According to Whang, alkaline water is drinkable at any time. He explains that there is always a net gain of alkalinity in the body when we drink alkaline water. This is why it doesn't make any difference whether one drinks alkaline water on an empty stomach or a full stomach, he said in *Article 20*.

Don't drink the **acid water**, that stuff is for use on the skin only. It acts just like perspiration (which is acid) and prevents any bacteria or viruses from entering the body. When it dries, as does sweat, it leaves a thin layer on the pores for protection. **Acid water** is for the outside of the body. It will act as an antiseptic on a wound; a treatment for a rash on the scalp or elsewhere; or it will stop bleeding. Saliva, also acidic, does the same. **Acid water** kills bacteria on contact. It is good for acne, cuts, scrapes, and rashes of all kinds.

Another piece of wisdom from McCauley's book is a quote from "*Ionized Water Explained*" by Dr Hidemitsu Hyashi of the Water Institute, Tokyo, who said :

Alkaline water, having a pH of between 9 and 11, will neutralize harmful stored wastes, and if you consume it every day it will gently remove them from your body. Yet, since the water is ionized, it will not leach out valuable minerals like calcium, magnesium, potassium, or sodium.

HOW MUCH ALKALINE WATER A DAY?

If five only 10oz (330ml) glasses of alkaline water a day is needed to improve your health, what happens if you drink more than that? Whang, quoting mind-numbing numbers of hydroxyl ions and H_2O molecules, said the first four glasses would cancel the acidic waste produced that day and the fifth glass would go towards reducing old accumulated wastes. He said you cannot hurry nature : any extra alkaline water would be wasted. You can, however, add a drop or two of Alkalife to increase the pH in the glass if you wished. This would help.

KILLING GERMS

The Japanese government wanted to know the germ-killing potential of alkaline and **acid waters**. They added bacilli (bacteria from the lining of the colon) to ordinary tap water, also to alkaline water and **acid water**. Here are the bacteria counts before and after ionization :

	LIVE BACTERIA COUNT
Original tap water before ionization	1400
Alkaline water immediately after	120
Alkaline water one hour after	0
Acid water immediately after	0
Acid water one hour after	0

They concluded that bacteria cannot live in either strong alkaline or **acid water**. Plant cells love **acid water**. When human cells are starved of oxygen due to an overload of acid, they use fermentation for oxygen even though they receive only ten percent of it. The cells don't die, they become acidic, and cancer looms. Plants take in carbon dioxide which is acidic and turn out oxygen which is why we get a huge oxygen boost when standing in a tropical rainforest or at a beach. **Acid water** stops insects from damaging plants. Rain water is better for plants than tap water as it is more acidic than tap water. If you cut flowers they will last longer in **acid water**. If you want those flowers to open up fully before the guests arrive, give them alkaline water. Animals age faster by giving them **acid wat**er and so do people, especially those who drink a lot of fizzy drinks.

HUMAN SKIN

Human skin is acidic, and we use alkaline soap on our skin to give it a good clean under the shower. But this can leave the skin open to germs. If, after your shower, you have some **acid water** from your filter available, douse it over your head, trunk and limbs. It can give your hair a conditioning job as well as stopping the germs. **Acid water** on your face will protect it from acne and discolouration. Most cosmetics have acid bases that match the pH of your skin. Dishwashing detergent commercials are true when they say they will make your hands smoother. Athlete's foot or haemorrhoids? Soak them in warm **acid water** for twenty minutes. You can soothe insect bites and neutralize their poison; put it on burns, in fact, do anything you like with **acid wat**er providing it is only on your outside skin.

COOKING

Alkaline water is good for making tea or coffee, mixing fruit juices, or making ice. Any water that will finish up inside your body should be alkaline water. Cooking rice in alkaline water will neutralize phosphorus acid with calcium before you eat the rice, cutting the amount of phosphoric acid the rice will develop later in your body. Incidentally, this applies also to cooked Basmati rice which does not stick together like normal rice and is healthier.

Boiling alkaline water will lower its pH because carbon dioxide from the air will be mixed in the water under heat. However, boiling water cannot boil away alkaline minerals; the minerals are all there combined with the bicarbonates. The benefits of neutralizing acids in your body will all be there after the volatile carbonic acid is reduced to water because you are breathing out the carbon dioxide.

WHISKY :

Whisky everywhere has a pH of 5.2 which is one-part acidic. Mix it with alkaline water one part to one, and the pH goes up to 9.5. Alkaline water with whisky tastes mild. The

Japanese claim you never get hangovers from drinking alkaline drinks.

*

CHAPTER 9

FAR INFRARED TECHNOLOGY

A scientific revolution started in Japan about 20 years ago. Far infrared (FIR) wave technology is taking over the traditional electrical fields of heating, drying, preservation, and healing. In the field of cooking, it is replacing the microwave oven. Microwaves have had their day, but they are potentially harmful to humans. To check this statement, take two healthy identical pot plants. Every day give one of them water from the tap and to the other water that has been brought to boiling point in your microwave and allowed to cool. After about nine days of watering, the plant getting the microwaved water will wilt and perhaps die; the other will continue to survive and thrive. Microwaving corrupts the DNA in a plant : the plant cannot recognise it. The same happens in our bodies.[18]

*

WHAT IS FAR INFRARED (FIR) TECHNOLOGY?

The use of far infrared wave therapy (FIR) has a significant connection with the medical field, especially when it is used to eliminate acid from the body. All energy from the sun comes in waves. The waves that microwaves use are of 30mm size; those of far infrared heating vary from 0.76mm to a size in microns. One micron is one-thousandth of a millimetre (mm). One far infrared wave therefore is about one thirty-thousandth the size of one microwave.

Far infrared waves are the same as those received and sent out by humans. They are therefore quite harmless to humans and other living creatures and are in fact beneficial. The FIR wave penetrates our bodies to a distance of about 7cm (2 ½") and the warming effect is very uniform. In comparison, heating by traditional electrical methods provides the surface with a hotter temperature so that our insides can be warmed by convection. The FIR wave is quicker, more uniform, and uses less power.

*

HEATING

A far infrared (FIR) sleeping pad or blanket is similar to an electric blanket in that you put it above a firm mattress and below the bottom sheet. You can adjust the temperature to your liking. The big difference of the FIR sleeping pad is that it heats you on the inside to melt away those acidic wastes which include fat wastes, and you therefore lose weight. You are doing the same as drinking alkaline water; but the big bonus is you don't have to do a thing except sleep, and everyone does that! The only change is that you might or might not have night sweats. Be happy with that, it's the acid wastes and fats leaving your body! The more alkaline water you drink the more you might sweat. If you don't sweat, you are probably losing your acid from urine, but the body seems to prefer perspiration to urine in getting rid of the acid. Remember, perspiration is acidic and acts to kill viruses and bacteria on the skin.

FIR waves are generated by passing a DC current through a carbon impregnated cotton sheet. The surface of the sleeping pad is not warm to the touch, but when you lie down on it you feel warm on the inside.

At first, one can sometimes experience dizziness when waking. This is similar to the feeling you might get when taking a hot bath. It is actually a sign that the pad is working. (See Appendix for more details on FIR pads, blankets and heaters.)

OVERHEAD HEATERS

Overhead FIR heaters are also used to good effect. The instant the heaters are turned on, you feel the warmth, but the air temperature remains cool. This saves energy and also gives you uniform heating.

Overhead FIR heaters can give the same temperature, say $26^0C/80^0F$ in a room for both summer and winter. In winter you may want to wear a jumper, whereas in summer just shirt sleeves. Why? Because although the room has the same air temperature, the walls and ceiling radiate more FIR waves in the summer than winter and that warms our bodies more.

FIR SAUNAS

Saunas heat your body and cause you to perspire. The conventional sauna heats the air which in turn heats your body. You need a fairly high temperature in the conventional sauna to do this because the heat rays are blocked in part by oxygen and nitrogen molecules. The FIR wave needs to provide only 42^0C and is not be blocked by the molecules. It will make you quite warm and to sweat profusely; and the air is not hot enough to burn your skin.

COOKING

You can boil an egg in a FIR boiler and the boiled egg can remain bacteria free for a month at room temperature. An FIR oven bakes bread with bubble sizes uniform throughout the bread. When foods are cooked in a conventional oven or microwave, part of the food is hot and the rest of it is heated by conduction. This can leave the food damaged or cooked unevenly. In a FIR oven the food is cooked uniformly and it does not damage the vitamins, minerals or flavourings with excessive heat.

WATERLESS EGG BOILER

Eggs do not need the 100^0C temperature of boiling water to be cooked. They cook fully in less heat although a little longer. FIR waves cook eggs without water uniformly and without damaging nutrients. After cooking, the eggs will not spoil for 30 days at room temperature.

HEALING

Science comes to the aid of medical healers with FIR therapy. In Chapter 5 we said that exercise had the benefit of heating the body's interior including the muscles. Sweat then took out the acid wastes. However, over-exercising is not recommended as it causes pools of acid to accumulate which in turn clogs the capillaries making removal of the wastes difficult.

FIR pads are good for deep heating for any area made painful from over-exercising. This expands the clogged capillaries without creating additional acidic wastes, and the wastes are removed more easily. Japanese doctors have reported that far infrared waves have been of help in treating a wide list of problems including stress induced chronic diarrhoea; numbness; shoulder, back and knee pain; low blood pressure; diabetes; radiation exposure related diseases; asthma; etc.

Dr Yoshiko Yamazaki in her book *The Science of Far Infrared Therapies* said the successes of FIR wave treatments have been due to the capability of the waves to remove toxins from the body. In general, toxins in our body are acidic. It follows that the FIR waves expand clogged capillary vessels and successfully dissolve hidden toxins into the blood and eventually out of the body via perspiration or urine.[19]

Whang agreed with Dr Yamazaki in saying he believed too much emphasis today has been on the conventional treatment of drugs to "destroy" the toxins at any cost. "*Usually, the cost is our own immune system,*" Whang said.

Yamazaki reported on a patient who had cancer of the esophageal orifice which would normally have killed him in a month and half. The cancer was caused by his longtime intake of an aspirin family pain killer which had poisoned his body. Far infrared therapy in a sauna excreted the accumulated poison, and the patient was discharged from the clinic after a stay of three months.

*

CHAPTER 10

OTHER THEORIES OF AGEING

Many are those who seek the Fountain of Youth. An age-old search going back to antiquity, the "fountain" has been discovered many times only to be proven false. The science of gerontology has grown significantly in the past 50 years since Leonard Hayflick in 1961 announced that blood cells divided a certain number of times (40 to 60) after which the organism slid down to its timely end. It was called The Hayflick Limit.

Hayflick was finally proved partly wrong, funds were provided, and the race was on for scientists to discover the answer to the quest for a long and healthful age. There were announcements of cellular enzymes, fibrobasts, and in-built clocks keeping time for the ageing process. Roy Walford showed that hungry young mice fed a highly nutritious diet containing 40 per cent fewer calories than normal could double their lifespans. Walford also displayed how lowering human body temperature a couple of degrees could enable us to live a little longer.

When a principle of thought suits a majority of people it is generally called a paradigm. Many paradigms about longevity have had their day in the sun, and when ultimately proven false they have usually been replaced with another better and equally false paradigm. Perhaps readers of this book may wonder if alkaline water--or rather the boosting of bicarbonate in the blood with alkaline water to give it its proper function--will be a lasting paradigm in the future, or not. The evidence so far from people like Warburg, Koch, Whang and, I expect, not a few Korean and Japanese researchers really does point that way. It certainly points that way for me.

Average human lifespan continues to increase over the years. Knowledge of the body and updated remedies to fix its diseases will keep pace. Do we wait and see, or do we do something and hope it's the winner? If on your deathbed you have done nothing to help yourself, perhaps you might be a little regretful when you see someone in the bed next to you, thirty years your senior, perhaps a centurian, with a bottle of alkaline water at hand and still joking with the nurses and asking to be sent home. Why wait for someone else to prove things conclusively, absolutely, convincingly, before you try it?

The use of alkaline water is gaining popularity as the word is spread. Intelligent responsible citizens continue to publish their opinions.

Ray Kurzweil, author of *Fantastic Journey* and one of the world's leading inventors, thinkers, and futurists said :

> *Ionized water is one of the simpliest and most powerful things you can do to combat a wide range of disease processes.*[20]

Dr Howard Peiper adds :

> *People who are waking up are taking responsibility for their own health. They are beginning to realize that the medical establishment is dependent on its human victims for its profits.*[21]

Dr Theodore A. Baroody declares :

> *I have administered over 10,000 gallons of this water* (to patients) *for about every health situation imaginable. I feel that restricted alkaline water can benefit everyone.*[22]

Dr George W. Crile of Cleveland, USA, one of the world's greatest surgeons, said :

> *There is no natural death. All deaths from so-called natural causes are merely the end-point of a progressive acid saturation.*[23]

*

CONCLUSION

Good health is a powerful force. When good health increases, the will to stay alive increases and the brain improves. It seems alkaline water, alkaline food in sensible quantities (mainly fruit and vegetables), far infrared therapy, and healthy living will enable the human species to survive and prosper to a ripe old age.

As far as the ionizer goes, let Bob McCauley, author of T*he Miraculous Properties of Ionized Water*, speak for all those who have discovered it :

Ionized water is one of the most significant preventative health advances of our generation and one of the most beneficial substances available to the human body. In importance, the invention of the water ionizer ranks with the great achievements of the 20th Century along with man first walking on the moon and the advent of the personal computer.

*

So here's to a long, healthy life and alkaline water, far infrared waves, and a good diet. Cheers!

APPENDIX

CHEMICAL EQUATION
Here, for the chemical-minded, is explanation of the equation in Chapter 1
$$H_2O + CO_2 + NaCl = HCl + NaHCO_3.$$
(Water + Carbon Dioxide + Sodium Salt = Hydrochloric Acid + SodiumBicarbonate.)
      The smallest thing on this planet is an atom. It has an electrical presence in the form of ions which circle around it like moons around a planet.
      An atom can be rock, soil, water, air, people, anything.
      Two or more atoms (they can be any number) make a molecule.
      Two or more molecules together can become a mineral or an element and can be changed in structure into other molecules if they are split or joined together.
      As you can see, the chemical equation above is made up of two combinations: one of three molecules and one of two.
      The first combination ($H_2 0 + CO_2 + NaCl$) contains water (two hydrogen & one oxygen atoms) + carbon dioxide (one carbon & two oxygen atoms) +

sodium salt (one sodium & one chloride atoms).   Total 8 atoms in three molecules.

The second combination (HCl + NaHCO$_3$) contains hydrochloric acid (one hydrogen & one chloride atoms) + sodium bicarbonate (one sodium & one hydrogen & one carbon & 3 oxygen atoms).   Total 8 atoms in two molecules.

Notice that the eight atoms in each combination are the very same atoms; the first combination has just been mixed up in the pot and poured out as something else.

TESTING THE pH OF URINE

Use ph Test Strips, see info@alkalinecookbook.com.au for details or use the same pH tester with its drops which came with your ionizer. Put 5ml urine in container, add drops provided, shake gently and check colour chart. Urine should not be above 7.0 (neutral). Below pH7 is normal. Clean container thoroughly after use. When testing for your urine's pH, do it after fasting for at least 12 hours or alternatively when you get up in the morning and before you exercise or eat or drink anything.

FAR INFRARED TECHNOLOGY

FIR products have not yet made a significant presence in Australia, but several items are available in Asia or the USA. I am sorry I cannot help but suggest you enquire via the Internet where there are details. People travelling overseas will probably find what they need in Asia at reasonable prices. FIR products are produced mainly in Korea, Japan, and Taiwan.

ORDERING ALKALIFE DROPS

These drops are patented and readily available in the U.S. Delivery by airmail takes 15 days from your email. Payment by credit card or bank transfer. Web site : www.alkalife.com.   Email to : info@alkalife.com.   Postal address :   Alkalife International, 8888 2W 129th Terrace, Miami, FL 33176-5945, USA.Phone 1-888-261-0870

ORDERING IONIZERS

Ionizers (alkaline filters) are made in Asia and also sold by Alkalife, USA. An Australian distributor is : Living Well Natural Health, 42 Sinclair Cres, Wentworth Falls, NSW 2782, ph 1300 330 701. The author uses a Chanson Miracle model made in Taiwan. If you have trouble fitting it, go to a Plumbers Supplies (see pink/yellow pages),and take the unit with you plus part of your tap fitting if possible. This may save you the expense of hiring a plumber.

ENDNOTES/BIBLIOGRAPHY

1   Professor Roy Walford, *Maximum Life Span* 1983. Norton, New York. p98

2   Sang Whang, *Reverse Aging* 1983. JSP Publishing, Miami.

3   Sang Whang, *Aging & Reverse Aging,* 2007. JSP Publishing,Miami.

4   Ibid. Sang Whang, *Aging & Reverse Aging*, p5

5   Internet: www.alternativemedicinedirect.com/AcidandAlkaline/excess-alkaline.php

6   Internet : http://alkalife.com/page.php?&cms=article*articleid=1. *Science & Health Series, 31 Articles*

7	Ibid. Walford

8	Arthur De Vany. *The New Evolution Diet* 2011. Vermilion, UK. Introduction.

9	Loren Cordain. *The Nutritional Characteristics of a Contemporary Diet,* based upon Paleolithic food groups. Journal of the American Nutraceutical Assocn. (5) 15-24.

10	Ibid. Sang Whang. Internet - Article 17 *Understanding Cancer.*

11	Siddharta Mukherjee. *The Emperor of All Maladies* 2011. Harper Collins, London.

12	Dr Ruth Cilento. *Heal Cancer* 1993. Hill of Content, Melbourne, p 187.

13	Ibid. Cilento, p 170

14	Ibid. Sang Whang. Internet Article 16 *Exercise & Health.*

15	Ibid. Cilento p 231

16	Bob McCauley. *The Miraculous Properties of Ionized Water* 2006. Spartan Enterprizes, USA. p58

17	Ibid. Sang Whang. Internet Article 10 - *Medical Science & Natural Science*

18	W. P. Kopp, Forensic Research, AREC Operations, Australia: www.lessemf/mw-stnds

19	Yoshiko Yamazaki. *The Science of Far Infrared Therapies* 1987. Man & History, Tokyo

20	Ibid. McCauley, p 47

21	Dr Howard Peiper. *Create a Miracle With Hexagonal Water* 2008. ATN Publishing p11

22	Theodore A Baroody. *Alkalize or Die* 1991. Holographic Health. p 96

23	Ibid. Sang Whang. Internet Article 20. Acidosis.

24	Ibid. McCauley, p46.

Why is it the most important discoveries seem to be the simple ones, discoveries that someone has just tripped over? Isaac Newton discovered the Law of Gravity when an apple fell on him; Archimedies found the important Law of Hydraulics when taking a bath.

Scientist/inventor Sang Whang suddenly tripped over the secret of preventing and in many cases curing disease when reading a Korean book of basic chemistry. The secret was the boosting of bicarbonates to eliminate acid build-up in our bodies and blood. And the way to do that was with alkaline water.

The discovery has enormous ramifications. This "magic water," made from ordinary tap water fed through a simple domestic ionizer or supplemented with a few drops of the chemicals potassium hydroxide and sodium hydroxide in the correct proportion (now patented) can retard or reverse ageing and lead to the prevention of cancer, diabetes, lung troubles, heart disease, and many of the so-called degenerative diseases now rife in the world.

In 1990, Korean/American Sang Whang wrote *Reverse Aging*, a book that gave us proof; quoted sound technical principles; and provided all the necessary data. But the only people impressed were ordinary people. The doctors, surgeons, hospitals, and drug companies turned a deaf ear or a blind eye. It was far too much of a seachange for them, or it may have threatened their incomes.

Alkaline water is now a regular intake for thousands of Japanese, Koreans, Chinese, and many other Asians. The longest-living industrialised people in the world are the Japanese. Alkaline water is fast becoming known and used in the USA and other English speaking countries.

Although Whang wrote in his most simple manner, he was a scientist and his books are full of technical information : a more simple book was needed. We hope it is the book you now have in your hands.

www.ingramcontent.com/pod-product-compliance
Lightning Source LLC
Chambersburg PA
CBHW080838170526
45158CB00009B/2582